ANIMAL MOVEMENTS

JEMI RAJA

To order additional copies of this book, contact
Toll Free 800 101 2657 (Singapore)
Toll Free 1 800 81 7340 (Malaysia)
www.partridgepublishing.com/singapore
orders.singapore@partridgepublishing.com

ISBN
ISBN: 978-1-5437-5421-6 (sc)
ISBN: 978-1-5437-5422-3 (e)

Print information available on the last page.

09/17/2019

PARTRIDGE

GATHERING INSPIRATION FROM GOD ABOVE AND FAMILY'S UNWAVERING SUPPORT. THANKS SO MUCH.

WITH A GRATEFUL HEART, DEDICATE THIS BOOK TO MY DEAR GRAND NEPHEWS AND NIECES

THIS BOOK BELONGS

TO

____

____

FROM

____

____

Birds fly.

Butterfly
flutters by.

Elephant
sways left
and right.

Tortoise and snail crawl with all might.

Crab and turtle can crawl and swim.

Whereas
the fish
only swims.

Creepy
crawler is
the insect
that creeps.

Turkey
wobbles
and worms
wriggle.

Horse gallops and rabbit hops.

What do you think other animals do?

Think

Now, how about you?

THE END

Printed in the United States
By Bookmasters